Number Theory's Unseen Connections

Unraveling the intricate tapestry of integers, rational numbers, and the beautiful complexity of modular forms

Norbert Quantock

Table of Contents

1 Introduction to Number Theory

Numbers, the seemingly simple building blocks of mathematics, hold a universe of intricate relationships waiting to be uncovered. This chapter serves as your gateway to the captivating world of number theory, a field brimming with puzzles and elegant solutions. We'll start with the basics, laying a solid foundation for your journey into this fascinating realm.

Understanding integers - positive, negative, and zero - is paramount. These are the numbers we'll be working with most frequently. Think of them as the fundamental characters in our mathematical story. Their properties, such as evenness and oddness, play a crucial role in many number-theoretical concepts. For example, the sum of two even numbers is always even, while the sum of an even and an odd number is always odd. These seemingly simple observations are the cornerstones upon which more complex theories are built.

Divisibility, a key concept, describes the relationship between two integers. If an integer a divides an integer b without leaving a remainder, we say a is a divisor of b, or b is divisible by a. This leads to the idea of factors, multiples, and prime numbers—numbers divisible only by one and themselves. The concept of divisibility underpins many calculations and proofs in number theory. Consider the number 12; its divisors are 1,

2, 3, 4, 6, and 12. Understanding divisibility forms the bedrock for more advanced concepts such as prime factorization.

Knowing divisibility rules speeds up calculations considerably. For instance, a number is divisible by 2 if its last digit is even (0, 2, 4, 6, or 8). A number is divisible by 3 if the sum of its digits is divisible by 3. A number is divisible by 5 if its last digit is either 0 or 5. These rules are not just shortcuts; they illustrate fundamental relationships between numbers and their digits. They're your secret weapons for quickly determining divisibility without resorting to lengthy calculations, providing a powerful tool for solving problems more efficiently. Mastering these rules will significantly enhance your ability to analyze and manipulate numbers.

Beyond these basic rules, the concept of the greatest common divisor (GCD) and the least common multiple (LCM) becomes crucial. The GCD is the largest number that divides both of two given integers without leaving a remainder. The LCM, on the other hand, is the smallest number that is a multiple of both integers. These concepts are instrumental in simplifying fractions and solving various problems involving ratios and proportions. Finding the GCD and LCM efficiently requires understanding prime factorization, a method to express a number as a product of its prime factors.

Prime factorization is arguably the most important concept in introductory number theory. Every integer greater than 1 can be uniquely expressed as a product of prime numbers. This fundamental theorem of arithmetic is the cornerstone of many advanced number-theoretic results. It's like having a unique fingerprint for each number, allowing us to categorize and analyze them in a structured way. Think of primes as the fundamental building blocks of all integers - just as atoms are the fundamental building blocks of matter. This unique representation unlocks solutions to many seemingly intractable problems.

Now, let's consider some simple applications. Imagine you are trying to divide two large numbers. Before you grab your calculator, consider the prime factorization of both. This approach simplifies division and allows you to find common factors quickly. The simplification process often reveals hidden relationships between numbers and facilitates the solution to various mathematical problems, such as solving equations or analyzing patterns. The ability to factor large numbers quickly is also critical in cryptography, which we'll explore later in the book. For now, focus on understanding the underlying principles and developing your skills in prime factorization.

This introduction has equipped you with fundamental tools. In upcoming chapters, we'll build upon these foundations, venturing into the intricacies of modular arithmetic, Diophantine equations, and the elegant world of modular forms. The journey ahead is filled with challenging problems and rewarding discoveries. Remember, the seemingly simple world of integers harbors a depth and complexity that will constantly surprise and amaze you. Embrace the challenge, and enjoy the exploration.

1.1 Basic Concepts

Numbers, the seemingly simple building blocks of mathematics, hold a universe of intricate relationships within themselves. We begin with integers, those whole numbers stretching infinitely in both positive and negative directions, including zero, the neutral point, the separator between positives and negatives. These integers, seemingly straightforward, form the basis of our exploration. They are the foundation upon which more complex concepts are built, the very essence of number theory itself. Understanding their fundamental properties is paramount to unlocking the deeper mysteries that lie ahead.

Consider divisibility. This seemingly simple concept—whether one integer can be divided evenly by another—opens a world of possibilities. Think of it as a test of compatibility between numbers. A number is divisible by another if the division results in a whole number, leaving no remainder. For instance, 12 is divisible by 3 because 12 divided by 3 equals 4, a whole number. However, 12 is not divisible by 5 as the division produces a remainder. This seemingly trivial observation becomes a crucial tool as we progress through more complex mathematical landscapes. It's the key that unlocks many doors to higher-level number theory concepts.

Prime numbers, those integers divisible only by 1 and themselves, are the fundamental building blocks of all other integers. They're like the atoms of the number world, indivisible and unique. Understanding primes is essential because every integer (excluding -1 and 1) can be expressed as a unique product of primes. This fundamental fact, known as the Fundamental Theorem of Arithmetic, is a cornerstone of number theory. This unique factorization is crucial; it's the framework upon which many complex theories are constructed. It provides a way to classify and categorize numbers, laying the foundation for exploring their relationships and properties. Think of trying to build something without knowing the basic materials—it's simply not possible.

Now, let's consider the concept of greatest common divisor (GCD), which finds the largest integer that divides both of two given integers without leaving a remainder. Imagine finding the largest possible square tile that can perfectly cover a rectangular floor without any gaps or overlaps—that's the essence of the GCD. It signifies a common factor, a shared property between two numbers. Finding the GCD is essential in simplifying fractions, and unexpectedly, plays a surprisingly vital role in certain cryptographic techniques. This seemingly simple idea has practical applications far beyond the theoretical.

Moving beyond individual integers, we encounter sets. A set is simply a collection of distinct objects—in our case, numbers. We might consider a set of even numbers, or a set of prime numbers, or even a set of numbers divisible by 7. These sets allow us to group numbers with shared properties, facilitating the analysis and comparison of these properties within larger structures. This organized approach is fundamental to understanding how various number types interact. The use of sets allows for a methodical approach to exploring the vast landscape of numbers, providing structure and order to what might otherwise feel chaotic.

Finally, let's briefly explore the idea of rational numbers. These are numbers that can be expressed as a ratio of two integers (a fraction), where the denominator is not zero. These encompass the integers themselves, but also introduce numbers like 1/2, 3/4, or -5/7. They're a broader category encompassing integers and many other numbers, extending the possibilities beyond the whole numbers. Their inclusion dramatically expands the mathematical landscape, enabling the exploration of new and complex relationships. Rational numbers form the stepping stone to even more intricate number systems, bridging the gap between the simple elegance of integers and the limitless expanse of real and complex numbers. This chapter provided a foundational overview of core concepts; the next chapters will delve deeper into the rich tapestry of number theory.

1.2 Divisibility Rules

Knowing divisibility rules is like having a secret shortcut through a mathematical maze. They let you quickly determine if a number is divisible by another without lengthy division. For instance, a number is divisible by 2 if its last digit is even—0, 2, 4, 6, or 8. This is because any even number can be expressed as 2 times another integer, meaning it's a multiple of 2. This simple rule saves countless seconds, especially when working with larger numbers.

Divisibility by 3 is equally elegant. A number is divisible by 3 if the sum of its digits is divisible by 3. Consider the number 123. Adding its digits (1 + 2 + 3 = 6), we get 6, which is divisible by 3. Therefore, 123 is divisible by 3. This rule works because of the modular arithmetic properties of our base-10 system; each power of ten can be written as 1 more than a multiple of 3 (10 = 33 + 1, 100 = 333 + 1, etc.). This rule is much faster than attempting a long division to find out the divisibility by 3.

Now let's tackle divisibility by 4. A number is divisible by 4 if its last two digits form a number divisible by 4. For example, 1312 is divisible by 4 because 12 is divisible by 4. This rule arises from the fact that 4 is 2 squared, so the number must be divisible by 2 twice. This means that the last two digits are crucial for this rule. Trying to divide by four a huge number can take a long time, but using this rule allows you to do the same operation much more efficiently.

The rule for 5 is perhaps the simplest. A number is divisible by 5 if its last digit is either 0 or 5. Think of the multiples of 5: 5, 10, 15, 20, and so on. They all end in 0 or 5. It's intuitive and easy to remember, making it a very practical rule. Quickly identifying the divisibility of a large number by 5 is essential and these rules make the task considerably easier.

Divisibility by 6 combines the rules for 2 and 3. A number is divisible by 6 only if it's divisible by both 2 and 3. This stems from the fact that 6 is the product of 2 and 3, therefore, you need to check if the last digit is even and if the sum of the digits is divisible by 3 to ensure the number's divisibility by 6. This avoids the need for a division in case you only need to find whether or not the number is divisible by 6.

Moving on to 8, a number is divisible by 8 if its last three digits form a number divisible by 8. This is a more complex rule as

8 equals 2 to the power of 3. For example, 12320 is divisible by 8 because 320 is divisible by 8. This again arises from the powers of 2, showing a relationship between divisibility by 2, 4, and 8. This is a quite complex rule, but still useful to save time and effort.
For divisibility by 9, there's a similar pattern to the rule for 3. A number is divisible by 9 if the sum of its digits is divisible by 9. For instance, 369: 3 + 6 + 9 = 18, and 18 is divisible by 9; therefore, 369 is divisible by 9. This property shares a similar origin with the divisibility rule for 3.
Finally, a number is divisible by 10 if its last digit is 0. It's the simplest of them all, a direct consequence of our decimal system. This rule is probably the one that everyone remembers and can instantly apply, and yet it is still useful and fundamental to know, especially for beginners.
Mastering these divisibility rules isn't just about memorization; it's about understanding the underlying mathematical principles. They highlight the structure of the number system, providing efficient ways to check for divisibility without tedious calculations. Practice makes perfect, so try applying these rules to various numbers. You'll quickly see how much time and effort they can save, making you a more confident and efficient mathematician. Remember, these rules are building blocks for more advanced concepts in number theory.

2 Prime Numbers and Factorization

Prime numbers, those enigmatic integers divisible only by one and themselves, form the bedrock of number theory. Understanding them is crucial; they're the fundamental building blocks from which all other integers are constructed. Think of them as the indivisible atoms of the number world. Their seemingly simple definition belies a deep complexity, a complexity mathematicians have grappled with for centuries, leading to beautiful and powerful theorems.

The Sieve of Eratosthenes offers a surprisingly simple, yet elegant, method for identifying primes. Imagine a grid of numbers; you systematically eliminate multiples, starting with the smallest prime, two. What's left? Primes! It's a visual and intuitive approach, perfect for beginners to grasp the concept of prime numbers. This method provides a practical way to find primes up to a certain limit. It's not the most efficient for extremely large numbers, but for learning, it works perfectly.

Now, consider the Fundamental Theorem of Arithmetic. This theorem states that every integer greater than one can be expressed as a unique product of primes. This seemingly simple statement is profound. It means that any number, no matter how large or seemingly complex, can be broken down into its prime constituents. Think about it: a seemingly random number, like 1,234,567, ultimately boils down to a unique combination of prime numbers. This theorem makes many mathematical processes far simpler, and acts as a fundamental building block in much of what comes later.

Let's explore a significant implication of the Fundamental Theorem: factorization. Factoring a large number into its prime components is, surprisingly, a computationally hard problem. This difficulty forms the basis of many modern cryptography systems. Secure online transactions, for example, rely on the time it takes to factor massive numbers. While easily understood in concept, the practical application demonstrates the surprising power of prime numbers.

The Prime Number Theorem provides a fascinating glimpse into the distribution of primes. While there's no simple formula to predict the next prime, this theorem gives us an estimate of how many primes exist up to a given number. It's not about finding specific primes; instead, it's about understanding their overall distribution. It's a statement about the 'density' of primes among all integers, a sophisticated piece of mathematical insight and a testament to the beauty and complexity of prime numbers. The theorem highlights the asymptotic nature of prime distribution, a concept that can be challenging but is extremely important to understand the behavior of primes in the long run. It shows that even though primes seem scattered, their distribution follows a predictable pattern.

Now, you might wonder why this matters. Beyond the intellectual satisfaction of unraveling these mathematical mysteries, prime numbers and factorization are foundational to many areas of modern technology. Consider RSA encryption, the system that protects sensitive information online. RSA's strength relies on the difficulty of factoring large numbers. This directly relates to the challenges and subtleties of working with prime numbers. Understanding primes is not just an abstract mathematical pursuit; it's a crucial component of the digital world we inhabit.

However, the study of prime numbers extends far beyond cryptography. In areas like coding theory, efficient methods of prime factorization are critical to creating effective error-correcting codes. Such codes are essential in ensuring reliable data transmission, from satellite communications to everyday internet browsing. The more we understand primes, the more robust our communication systems become. The applications of prime numbers constantly evolve, highlighting their lasting significance.

Finally, remember that this chapter is just the beginning. We've touched upon the surface of a rich and fascinating area of mathematics. The world of prime numbers is full of unsolved problems and ongoing research. Your journey into number theory has just begun, and the exploration of primes is a vital part of that journey. The more you investigate, the more amazing the intricate tapestry of numbers will become.

2.1 Sieve of Eratosthenes

Imagine a vast grid, stretching endlessly. Each cell holds a number, starting from two. This is the heart of the Sieve of Eratosthenes, a surprisingly simple yet powerful method for finding prime numbers, those elusive integers divisible only by one and themselves. It's a beautiful visual representation of a fundamental concept in number theory, a field dedicated to exploring the intricate relationships between numbers. Think of it as a treasure hunt, where prime numbers are the hidden gems. We begin by circling the number two, our first prime. Then, we systematically eliminate every multiple of two - four, six, eight, and so on. We systematically eliminate all multiples of two. They're not prime because they're divisible by something other than themselves and one.

Now, consider the next uncircled number: three. This too, is prime. Next, we eliminate all multiples of three—six (already eliminated), nine, twelve, fifteen, and so forth. Observe how the process systematically removes composite numbers, leaving the primes untouched. We continue this process, circling the next uncrossed number (five in this case) and then eliminating its multiples. Each time, we're left with a progressively refined set, honing in on the primes. It's almost like refining gold from ore - systematically removing impurities to reveal the precious metal. This methodical approach is surprisingly efficient, especially for identifying primes within a reasonable range. You'll find that as you proceed, you need only eliminate multiples of primes up to the square root of your highest number. Why? Because any composite number larger than that will already have been crossed out as a multiple of a smaller prime.

Let's illustrate this with a specific example. Let's say we want to find all primes up to 50. We begin by creating a grid of numbers from 2 to 50. First, we circle 2 and cross out all multiples of 2. Then, we move to the next uncrossed number, 3, circle it, and cross out all multiples of 3. We continue this process with 5, 7, 11, 13, until we reach the square root of 50, which is approximately 7.07. This means that all remaining uncrossed numbers are prime. You'll have a simple, yet effective way to list out all primes less than 50, a process that would otherwise involve many trial divisions. The elegance of this method lies in its simplicity and efficiency. This iterative process reveals the primes hidden within the seemingly chaotic arrangement of integers. It's a fascinating illustration of how order can emerge from apparent randomness.

The Sieve of Eratosthenes is more than just a method for finding primes; it's a gateway to understanding the fundamental nature of numbers. It's a technique that allows us to visualize the structure of prime numbers and get a glimpse of their distribution. It offers a tangible method to understand what it means for a number to be prime and what makes a number composite. It allows for a deeper appreciation of prime numbers. Consider that primes act as the fundamental building blocks of all other integers, just like atoms form the basis of all matter. The Sieve provides a hands-on experience to appreciate this vital role in number theory. Every composite number is uniquely constructible by multiplying primes. Furthermore, the algorithm underlying the Sieve is surprisingly efficient for finding primes within a specific range. The computational complexity, in terms of the number of operations required, scales relatively well, making it a practical tool for relatively large ranges. This makes it a computationally inexpensive way to sift for primes, a feat achieved using very simple operations which makes it more efficient than other methods. This efficiency is a testament to the elegance of the algorithm itself, a product of its clever design. In other words, the method minimizes redundancy, a highly sought-after trait in algorithm design. This underlying elegance is a fascinating area to explore, bridging the gap between the mathematical theory and its efficient algorithmic implementation. It's a perfect example of how theoretical concepts can lead to practical and efficient computational methods. The Sieve of Eratosthenes is not just a mathematical curiosity; it's a foundational tool that underscores the beauty and power of number theory and its importance in numerous areas of mathematics and computer science. It's a testament to the inherent elegance and utility found within seemingly simple mathematical procedures. It's a method that remains relevant even in today's world of advanced computing. Therefore, understanding the Sieve is a crucial step in further exploration of prime numbers and number theory in general, highlighting the connection between theoretical concepts and practical applications. The algorithm's efficiency makes it a valuable tool even in the age of powerful computers. It showcases the enduring relevance of classical mathematical methods. You can even try to implement this algorithm in a programming language as a fun exercise.

2.2 Fundamental Theorem

Every positive integer greater than 1 can be expressed as a unique product of prime numbers, except for the order of the factors. This seemingly simple statement, the Fundamental Theorem of Arithmetic, is a cornerstone of number theory. It's not just a statement; it's a powerful tool that unlocks deeper understanding and allows us to approach more complex problems systematically. Think of prime numbers as the fundamental building blocks of all integers, like atoms forming molecules. Just as we can describe a complex molecule by listing its constituent atoms and their arrangement, we can uniquely represent any integer by its prime factorization. This uniqueness is critical; it assures us there's only one way to break down a number into primes, ignoring the order.
Consider the number 12. We can factor it as 2 x 2 x 3 or 2 x 3 x 2 or 3 x 2 x 2; all represent the same prime factorization, which is often written as 2^2 x 3. Similarly, 70 becomes 2 x 5 x 7. This might seem obvious for small numbers, but the theorem guarantees this uniqueness holds true even for astronomically large integers, numbers far exceeding anything we can practically write down. This unique decomposition is what makes the theorem so significant; it's the key to numerous mathematical techniques and applications.
Understanding the Fundamental Theorem helps us solve various problems involving integers. For example, finding the greatest common divisor (GCD) of two numbers becomes easier. Instead of brute-force methods, we can use prime factorization. Let's say we want to find the GCD of 12 and 18. Factoring 12 gives 2^2 x 3, and 18 factors into 2 x 3^2. The GCD is found by taking the minimum power of each common prime factor. In this case, the common prime factors are 2 and 3, with minimum powers of 2^1 and 3^1. Therefore, the GCD of 12 and 18 is 2 x 3 = 6. This approach is far more efficient

for larger numbers than other methods, proving the power of understanding the prime factorization.

The theorem's power also extends to the realm of least common multiples (LCMs). Finding the LCM of two or more numbers is simplified by using prime factorizations. Let's find the LCM of 12 and 18 again. From their prime factorizations, 2^2 x 3 and 2 x 3^2, we choose the maximum power of each prime factor present in either factorization. This results in 2^2 x 3^2 = 36, which is the LCM of 12 and 18. This method provides a systematic way to find the LCM even for large sets of numbers, a task challenging using other methods.

Beyond GCD and LCM calculations, the Fundamental Theorem underpins many advanced concepts in number theory. It's fundamental to the study of modular arithmetic, a system where integers are treated according to their remainders when divided by a given modulus. In cryptography, a field heavily reliant on number theory, the unique prime factorization of integers is crucial for designing secure encryption algorithms. For example, the RSA algorithm, widely used in secure online transactions, depends on the difficulty of factoring large numbers into their prime factors. The larger the numbers, the more computationally intensive factorization becomes, making these algorithms secure. The Fundamental Theorem ensures that the security of these algorithms is based on a solid mathematical foundation.

The seemingly simple idea of unique prime factorization has far-reaching consequences. It's not just about breaking down numbers; it's about understanding their fundamental structure and how this structure allows us to solve problems and build more complex mathematical structures. The theorem is a stepping stone to more advanced concepts, making it an essential pillar of further explorations in number theory. Mastering its implications provides a strong foundation for future study of more sophisticated and intricate topics in the field. While it might seem like a straightforward concept, appreciating the full implications of the Fundamental Theorem of Arithmetic is vital for anyone seeking a deeper understanding of the fascinating world of numbers. Its elegant simplicity belies the profound impact it has had on mathematical progress and future developments. The theorem's power continues to inspire new discoveries and applications, further solidifying its position as a core element in number theory. Therefore, a thorough understanding of this theorem is essential for any aspiring mathematician or anyone captivated by the beauty and logic of numbers.

2.3 Prime Number Theorem

Imagine a vast, seemingly endless field of numbers. Scattered throughout are the primes, those enigmatic integers divisible only by one and themselves. Understanding their distribution, seemingly random yet governed by profound patterns, is crucial in number theory. This chapter dives into the Prime Number Theorem, a landmark result that unveils a surprising regularity amidst this prime number chaos. It doesn't tell us exactly where each prime lies, but it reveals a remarkable connection between the number of primes less than a given number and a simple function.

The Prime Number Theorem states that the number of primes less than or equal to x, denoted (x), is approximately equal to x/ln(x), where ln(x) is the natural logarithm of x. As x gets larger, this approximation becomes increasingly accurate, suggesting a hidden order. Think of it like this: if you're searching for primes in a huge number field, this theorem provides a rough estimate of how many you should expect to find. It's not a precise count, but a powerful guideline, revealing the overall density of primes. It's a bridge between the seemingly random distribution of prime numbers and the predictability of mathematical functions, an elegant unification of the apparently disparate.

Let's unpack the intuition behind this theorem. The density of primes decreases as we move towards larger numbers. We don't encounter primes as frequently as we move towards infinity. The natural logarithm function, ln(x), captures this diminishing density; it grows slowly, reflecting the decreasing frequency of primes. The ratio x/ln(x) essentially adjusts the number x to account for this thinning out of primes, providing a remarkably accurate estimate of their count.

Proving this theorem is a significant feat in mathematical history. It required advanced tools and concepts, with early attempts falling short. The first rigorous proof came relatively recently, built upon the backs of various mathematicians and their contributions. This illustrates the collaborative nature of mathematical discoveries, emphasizing the building of ideas over time. The collaborative nature of mathematics, where individual contributions build towards grand theorems, mirrors scientific progress across disciplines. The theorem itself isn't easily proven, demonstrating how sometimes seemingly simple statements necessitate complex mathematical machinery for their justification.

However, the beauty of the Prime Number Theorem lies not just in its sophisticated proof, but in its wide-ranging applications. It forms the cornerstone of many advanced number theory concepts. It helps us analyze and understand the distribution of primes in various contexts, improving our ability to predict their prevalence within different intervals. Moreover, it has implications for cryptography, where prime numbers form the basis of secure encryption techniques. The theorem's ability to approximate the number of primes within a given range is crucial for various algorithms in the field. It facilitates the generation of large prime numbers used in secure communications, demonstrating the practical utility of abstract mathematical concepts.

Understanding the Prime Number Theorem provides you with

a deep appreciation of the intricate relationships within the world of numbers. It bridges the gap between the seemingly random distribution of primes and the elegance of mathematical functions. While the proof itself may involve advanced calculus and complex analysis, the fundamental idea—that we can accurately approximate the number of primes using a relatively simple formula—is both elegant and surprisingly intuitive. This understanding is crucial for moving towards advanced topics like cryptography and the study of modular forms, where the distribution and properties of primes play a significant role. It is a crucial element in your journey to mastering the beauty and power of number theory. This theorem, at its core, reflects the underlying harmony between apparent randomness and profound mathematical order.

3 Congruences and Modular Arithmetic

Imagine a clock. Its face shows only numbers from 1 to 12. Once the hour hand passes 12, it starts again at 1. That's modular arithmetic in a nutshell: a system where numbers "wrap around" after reaching a certain value, called the modulus. In our clock example, the modulus is 12. We write this as "a b (mod m)," meaning 'a' and 'b' leave the same remainder when divided by 'm'.

This seemingly simple concept unlocks a wealth of possibilities. Consider finding remainders. We might be interested in knowing what's left over when 47 is divided by 5. Instead of performing the division, we can use modular arithmetic. 47 2 (mod 5), since 47 divided by 5 leaves a remainder of 2. This simple method can be highly effective for handling large numbers. Using modular arithmetic, complex calculations become significantly easier.

Let's delve into linear congruences - equations of the form ax b (mod m), where we need to solve for 'x'. For example, 3x 2 (mod 5). We can solve this by trial and error, or use more sophisticated methods such as the extended Euclidean algorithm (explained elsewhere in the book). These equations, seemingly simple, appear in various branches of mathematics and computer science, from cryptography to coding theory.

Finding solutions in modular arithmetic requires understanding the properties of the modulus and the numbers involved. For instance, we might need to find the greatest common divisor (GCD) of two numbers to determine whether a solution even exists!

The Chinese Remainder Theorem is a remarkably elegant result. Suppose you have a system of congruences, like x 2 (mod 3) and x 3 (mod 5). The Chinese Remainder Theorem guarantees a unique solution for 'x' (within a certain range) if the moduli are pairwise coprime (meaning they share no common divisors other than 1). Imagine solving a series of interconnected puzzles simultaneously; that is essentially what the Chinese Remainder Theorem allows you to do. It's a powerful tool for solving seemingly impossible problems in various contexts.

Fermat's Little Theorem is a cornerstone of modular arithmetic. It states that if 'p' is a prime number, and 'a' is an integer not divisible by 'p', then a^(p-1) 1 (mod p). Think of it as a surprising property of prime numbers. This theorem has profound implications in number theory and forms the basis for several cryptographic algorithms, enabling secure communication in the digital world. This is a truly remarkable result connecting seemingly unrelated concepts.

Euler's Totient Theorem generalizes Fermat's Little Theorem. While Fermat's Little Theorem deals specifically with prime moduli, Euler's Totient Theorem extends the concept to any positive integer modulus. It involves the Euler totient function, (n), which counts the number of positive integers less than or equal to 'n' that are relatively prime to 'n'. The theorem states that if 'a' and 'n' are coprime, then a^(n) 1 (mod n). It's a powerful tool, used extensively in cryptography and other areas of number theory. Understanding the intricacies of this function and its connections to other parts of number theory allows us to solve problems of exceptional complexity.

Modular arithmetic might seem like an abstract concept, but

its applications are widespread. It's the backbone of public-key cryptography, ensuring secure online transactions. It helps in coding theory, enhancing error detection and correction in data transmission. Even some aspects of physics use modular arithmetic to model certain phenomena. The beauty of modular arithmetic lies in its simplicity and its power to elegantly solve complex problems in various fields of study. Mastering this concept is a step towards appreciating the elegant intricacy of number theory. It empowers you to simplify otherwise daunting calculations and unlocks a gateway to deeper explorations of prime numbers and cryptography. The seemingly simple clock analogy, remember, hides a deep and multifaceted world of mathematical wonders.

3.1 Modular Operations

Modular arithmetic is like a clock. Imagine a 12-hour clock; when you go past 12, you start again at 1. This "wrapping around" is the essence of modular arithmetic. We use a modulus, denoted as n, which determines when the "wrap-around" occurs. For example, if n = 12, then 13 is congruent to 1 (13 1 mod 12), 25 is congruent to 1 (25 1 mod 12), and so on. Understanding this fundamental concept unlocks a whole world of mathematical possibilities.

The basic operations - addition, subtraction, multiplication - are performed as usual, but the result is then reduced modulo n. So, if we want to calculate 7 + 9 mod 12, we add 7 and 9 to get 16, and then find the remainder when 16 is divided by 12, which is 4. Thus, 7 + 9 4 mod 12. Subtraction and multiplication follow the same principle: the result is always reduced to its remainder when divided by the modulus. This simplicity hides surprising depth; these seemingly simple operations form the foundation of sophisticated mathematical concepts.

Division is a bit trickier in modular arithmetic. To divide a by b modulo n, we need to find an integer x such that bx a mod n. This isn't always possible. If the greatest common divisor (GCD) of b and n is greater than 1, there might be no solution. However, if GCD(b, n) = 1, meaning b and n are coprime, a unique solution exists. Finding this solution often involves using the Extended Euclidean Algorithm, a powerful tool for finding the greatest common divisor and expressing it as a linear combination of the two input numbers. This algorithm allows us to solve linear Diophantine equations, which are crucial in many areas of number theory and cryptography. Let's tackle an example. Suppose we want to solve 3x 5 mod 7. Here, the GCD(3, 7) = 1, so a solution exists. We can use trial and error, or we can employ the Extended Euclidean Algorithm. This algorithm systematically finds integers x and y such that 3x + 7y = GCD(3,7) = 1. One solution is x = 5 and y = -2, giving 3(5) + 7(-2) = 1. Now, multiplying this equation by 5, we get 15x + 35y = 5. Reducing this modulo 7 gives 15x 5 mod 7, which simplifies to x 5 mod 7. Therefore, x = 5 is a solution, and all solutions are of the form x 5 mod 7 (or x = 5 + 7k, where k is any integer).

Modular exponentiation, calculating ab mod n, is crucial in many applications, particularly cryptography. Naively calculating ab and then reducing modulo n is inefficient for large values of b. A more efficient method is exponentiation by squaring. This involves repeatedly squaring a and reducing modulo n at each step, only multiplying the necessary squared terms to obtain the final result. This method dramatically reduces the number of calculations needed. Understanding modular exponentiation is key for understanding modern cryptography algorithms. Many cryptographic protocols, such as RSA, rely heavily on this concept for their security. The efficiency of modular exponentiation is paramount in securing sensitive information.

Consider the RSA cryptosystem, a cornerstone of modern cryptography. Its security relies on the difficulty of factoring large composite numbers, and modular exponentiation plays a critical role in both encryption and decryption. Public key cryptography uses two keys: a public key for encryption and a private key for decryption. The public key is easily distributed, while the private key is kept secret. Modular exponentiation is central to both the encryption and decryption processes. By carefully choosing the modulus n (the product of two large prime numbers), the RSA algorithm provides a highly secure method of information exchange. The practicality of modular exponentiation is directly linked to the practicality of secure communication in our digital world.

The exploration of modular operations only scratches the surface of modular arithmetic's potential. From solving congruences to underpinning modern cryptography, its significance is undeniable. Mastering these fundamental operations is your first step to unraveling the deeper mysteries of number theory. This is just the beginning of your journey into a world of intricate patterns and unexpected connections. The elegance and power of modular arithmetic will reveal themselves as you continue to explore. Continued practice and exploration are key to gaining proficiency.

3.2 Linear Congruences

Solving linear congruences is like deciphering a secret code, a puzzle where numbers behave in a cyclical, rather than linear, fashion. Imagine a clock; when you go past 12, you start again at 1. Modular arithmetic mirrors this behavior. A linear congruence is an equation of the form ax b (mod m), where a, b, and m are integers, and " denotes congruence. The goal? Finding integer values of x that satisfy the equation.

Finding solutions isn't always straightforward. Sometimes, a solution doesn't exist. Other times, multiple solutions exist, all neatly spaced apart, like stepping stones along a path. Under-

standing the greatest common divisor (GCD) of 'a' and 'm' is crucial. If GCD(a, m) doesn't divide 'b', there are no solutions - our stepping stones disappear into the void.
But if GCD(a, m) does divide 'b', there's hope! The number of solutions is precisely GCD(a, m). For instance, if GCD(a, m) = 1, then we have a single unique solution modulo m. This single solution can be found using the extended Euclidean algorithm, a clever method that finds integers x and y such that ax + my = GCD(a, m). This equation may seem unrelated at first, but it cleverly reveals the solution to our congruence.
The extended Euclidean algorithm involves repeatedly applying the division algorithm. Start by expressing the larger number as a multiple of the smaller number plus a remainder. Then, substitute the remainder into the equation, continuing this process until you reach a remainder of zero. Working backward, you can express the GCD as a linear combination of a and m, magically revealing the solution to our linear congruence.
Let's consider a concrete example: 3x 5 (mod 7). Here, a = 3, b = 5, and m = 7. The GCD(3, 7) = 1, which divides 5, so solutions exist. Applying the extended Euclidean algorithm, we find that 3(5) + 7(-2) = 1. Multiplying by 5, we get 3(25) + 7(-10) = 5. This tells us that 3(25) 5 (mod 7). Since 25 4 (mod 7), our solution is x 4 (mod 7).
However, what if GCD(a, m) > 1? The approach needs slight adjustment. We still check if GCD(a, m) divides 'b'—if not, no solutions exist. If it does, we can simplify the congruence by dividing the entire equation by the GCD. This results in a new, simpler congruence with a GCD of 1, making it solvable using the extended Euclidean algorithm. We then need to account for the solutions lost during the simplification by carefully determining the remaining congruent solutions.
Consider 6x 12 (mod 18). Here, GCD(6, 18) = 6, which divides 12. Simplifying, we get x 2 (mod 3). The solutions are x 2, 5, 8, 11, 14, 17 (mod 18), six solutions in total, which is precisely

GCD(6,18). Each solution represents a distinct solution modulo 18, each separated by a distance of 3.
Linear congruences are more than just abstract mathematical concepts; they form the foundation for many practical applications. They show up in cryptography, providing the mathematical basis for secure communication. Understanding the principles of linear congruences allows you to grasp the core mechanics behind these cryptographic systems. It provides you with the tools needed to understand the underlying magic which secures your daily online transactions and communications. They're also involved in scheduling and numerous other practical areas relying on modular arithmetic.
Mastering linear congruences opens a door to more advanced number theory topics. The Chinese Remainder Theorem, for instance, builds upon the principles we've explored here, allowing you to solve systems of linear congruences. This is essential for deeper understanding and more sophisticated applications in diverse mathematical and computational fields. Therefore, focusing on fully comprehending this chapter is very important in order to get a proper background in number theory. Through hard work and understanding, the world of congruences will become an open book for you. You'll discover the elegant interplay between seemingly disparate parts of mathematics.

3.3 Chinese Remainder Theorem

Imagine you have several boxes of chocolates, each with a different number of chocolates. You want to know the total number of chocolates, but you only know how many chocolates are left in each box after you've already eaten some. The Chinese Remainder Theorem helps solve exactly this kind of problem, but with a mathematical twist! It deals with systems of congruences - essentially, remainders after division.

Let's say you have three boxes. Box A has a number of chocolates, 'x', which when divided by 5 leaves a remainder of 2 (x 2 (mod 5)). Box B, when its number of chocolates, 'x', is divided by 7, leaves a remainder of 3 (x 3 (mod 7)). Finally, Box C, with its chocolates 'x', leaves a remainder of 2 when divided by 11 (x 2 (mod 11)). The Chinese Remainder Theorem provides a systematic way to find 'x', the total number of chocolates initially. It might seem impossible at first, a chocolatey mathematical mystery. But with the right method, we can solve it.

This theorem isn't just about chocolates; it's a fundamental tool in number theory. It helps us solve problems where we have information about remainders after division by different numbers. Think of it as a puzzle where the pieces are remainders, and the solution is the original number. We can't directly see the total, but we can deduce it using the clues—the remainders—provided.

Solving a system of congruences using the Chinese Remainder Theorem requires a bit of mathematical maneuvering. First, find the least common multiple (LCM) of the moduli—the numbers we're dividing by (in our chocolate example, 5, 7, and 11). The LCM is crucial because it reflects the periodicity of the remainders. This LCM represents the cycle length within which the total number of chocolates 'x' must fall.

Next, we find what's called a 'particular solution' for each congruence. That means, we'll find a value of 'x' that satisfies each congruence individually. It's like finding one possible solution for each box independently, ignoring the others for now. Each particular solution is the starting point in our search for the final solution. These individual solutions help us account for the unique relationship of 'x' with each modulus.

The beauty of the Chinese Remainder Theorem lies in its ability to combine these particular solutions into a single, unified solution that satisfies all congruences simultaneously. This is done using a clever combination of modular arithmetic and careful consideration of the least common multiples and individual congruences, ensuring the overall solution conforms to the remainders from every box. This merging of solutions is where the power of the theorem truly shines.

Now, let's consider how to find the general solution. Once we have one solution that satisfies all congruences, we can find infinitely many others! This is because, due to the cyclic nature of modular arithmetic, every multiple of the LCM will have the same remainders as our initial solution. Adding any multiple of the LCM to our solution still works. This reveals an infinite family of solutions.

The theorem's power extends far beyond chocolate box problems. It has significant applications in various fields. In cryptography, it plays a crucial role in secure communication protocols. The Chinese Remainder Theorem helps ensure the integrity and confidentiality of sensitive data by providing a way to encode and decode messages using modular arithmetic. It's part of several advanced encryption techniques.

Moreover, in computer science, the theorem is valuable for solving specific algorithmic problems efficiently. In situations where large numbers are involved, the theorem's ability to break down complex computations into smaller, more manageable ones is invaluable. The theorem enables solving complex problems that might be computationally infeasible otherwise.

Furthermore, its applications extend even into areas like scheduling and assigning tasks in an optimal way. Think about optimizing timetables for various events or resource allocation. This seemingly theoretical concept allows us to solve real-world problems in a systematic way. For example, consider synchronizing the operation of multiple machines in a factory, where different components operate at different speeds; the theorem can provide insight into scheduling these operations efficiently.

Even though the calculations involved in applying the Chinese

Remainder Theorem might seem daunting at first, remember the underlying principle is straightforward: it's all about combining information about remainders to deduce the original number. Mastering the theorem is about learning the techniques, understanding the underlying logic, and seeing the connections to various applications. It's a testament to the elegance and power of number theory, showcasing how seemingly abstract concepts can have far-reaching practical uses. You may encounter more complex examples in your later studies, but understanding the basic principle and the fundamental steps is the key to unlocking its potential. With practice, you'll confidently navigate the intricate world of modular arithmetic and the power of the Chinese Remainder Theorem.

3.4 Fermat's Little Theorem

Imagine a world where large numbers, seemingly unwieldy, become surprisingly manageable. That's the power of modular arithmetic, and Fermat's Little Theorem is a cornerstone of this fascinating field. It offers a surprisingly simple yet powerful way to explore the remainders of numbers after division. This theorem, elegant in its brevity, unlocks doors to sophisticated mathematical concepts and real-world applications.

Let's begin with the basics. Modular arithmetic deals with remainders. When we divide a number by another, the remainder is what's left over. For example, 17 divided by 5 leaves a remainder of 2. In modular arithmetic, we write this as 17 2 (mod 5), meaning 17 is congruent to 2 modulo 5. The modulus (5 in this case) determines the "size" of our number system; numbers "wrap around" once they reach the modulus.

Now, for Fermat's Little Theorem: if p is a prime number, and a is an integer not divisible by p, then a raised to the power of p minus 1, is congruent to 1 modulo p. Written mathematically, ap-1 1 (mod p). It's a deceptively simple statement, yet it holds profound implications.

Consider an example: let's use the prime number 7 and the integer 3. According to the theorem, 3(7-1) 36 1 (mod 7). Let's check: 36 = 729. Dividing 729 by 7 gives a remainder of 1, confirming the theorem. This seemingly simple calculation opens up a whole new world of possibilities.

The beauty lies in its applicability. We can use this theorem to quickly determine remainders for very large numbers without needing extensive calculations. Imagine trying to find the remainder when 2100 is divided by 11. Direct computation is impractical, but Fermat's Little Theorem simplifies the process significantly. Since 11 is prime, we know 210 1 (mod 11). Therefore, 2100 = (210)10 110 1 (mod 11). The remainder is 1 - achieved with minimal effort.

Beyond computational efficiency, Fermat's Little Theorem has crucial applications in cryptography. Many modern encryption systems rely on its properties. For instance, the RSA algorithm, widely used for secure online communication, relies heavily on modular arithmetic and prime numbers, making Fermat's Little Theorem an essential tool in securing digital transactions.

You might wonder how this relates to other concepts we've covered. The fundamental theorem of arithmetic, which states that every integer greater than 1 can be uniquely expressed as a product of prime numbers, is crucial. It's a key underpinning for Fermat's Little Theorem, since the theorem's prime number condition is so essential. Furthermore, the Chinese Remainder Theorem, another essential piece of modular arithmetic, complements Fermat's Little Theorem. While Fermat's Little Theorem helps us understand remainders in single modulo systems, the Chinese Remainder Theorem allows us to solve systems of congruences simultaneously, combining different moduli.

Understanding Fermat's Little Theorem isn't just about memorizing a formula; it's about appreciating its elegant simplicity and its far-reaching consequences. It's a testament to the interconnectedness of seemingly disparate areas within number theory, linking fundamental properties of prime numbers to powerful applications in modern technology. The theorem's power lies not only in its computational efficiency but also in its role as a building block for more advanced concepts in number theory and cryptography. Mastering this theorem is a significant step towards a deeper understanding of the intricacies of the number systems and their profound implications. It's an essential tool for anyone seeking a firm grasp of the unseen connections that unite the seemingly disparate areas within number theory. With practice and further exploration, you will discover even more profound applications of this seemingly simple yet remarkable theorem.

3.5 Euler's Totient Theorem

Imagine a lock with a complex combination. You need a key - a special number - to unlock it. Euler's Totient Theorem provides that key, offering a powerful shortcut in modular arithmetic. It tells us about the number of integers, less than a given integer n, that are relatively prime to n. These relatively prime numbers are those that share no common factors other than 1 with n. That's the core idea, and it's surprisingly useful. This theorem isn't just abstract theory; it has real-world applications. Consider cryptography, where secure communication hinges on complex mathematical operations. Euler's Totient Theorem is a cornerstone of RSA encryption, a widely used algorithm securing online transactions and sensitive data. It's like a secret code, enabling secure transmission of information. Understanding it helps you grasp the fundamental mathematics behind online security. The more you understand the core mathematical principles, the more you appreciate the technology safeguarding our digital lives.

Let's delve into the mathematical details. For an integer n,

Euler's totient function, denoted as (n), counts the positive integers up to n that are relatively prime to n. Calculating (n) directly can be tedious for large n, however, there are methods to simplify this task, especially if you know the prime factorization of n. If n is prime, (n) = n \item 1, which is quite straightforward. If n is a product of distinct primes, p \emph{ p } ... \emph{ p, (n) = (p - 1)(p - 1)...(p} - 1). For composite numbers with repeated prime factors, the calculation is slightly more involved, but still manageable.

Now, the theorem itself states that if a is an integer relatively prime to n, then a(n) 1 (mod n). This congruence means that a raised to the power of (n) leaves a remainder of 1 when divided by n. It's a powerful statement, offering a crucial relationship between exponentiation and modular arithmetic. Think of it as a cyclical property of numbers within a modular system. This seemingly simple statement underpins much of modern cryptography.

Let's work through a simple example. Consider n = 10. The integers relatively prime to 10 are 1, 3, 7, and 9. Thus, (10) = 4. Let's choose a = 3, which is relatively prime to 10. According to Euler's Totient Theorem, 34 1 (mod 10). Let's verify: 34 = 81, and 81 divided by 10 leaves a remainder of 1. The theorem holds true! This simple example illustrates the power and elegance of the theorem. Trying it with different numbers and observing the pattern is a great way to cement your understanding.

But why is this useful? One key application lies in cryptography. In RSA encryption, the security relies on the difficulty of factoring large numbers. The public key, used for encrypting messages, involves the product of two large prime numbers. The private key, necessary for decryption, relies heavily on Euler's Totient Theorem to calculate the inverse needed for decoding. This inverse is the mathematical 'key' to unlock the encrypted message, making Euler's theorem crucial for ensuring secure communication.

The beauty of Euler's Totient Theorem lies in its simplicity and profound implications. It connects seemingly disparate concepts in number theory—divisibility, prime numbers, and modular arithmetic—creating a powerful tool for solving complex problems. Mastering this theorem opens doors to understanding more advanced concepts in number theory and its diverse applications in fields like cryptography and computer science. Don't be intimidated by the mathematical notation. The key is to understand the underlying concepts and practice with different examples. This theorem is more accessible than you might initially think. Take your time, work through the examples, and soon you'll be using this elegant tool to unlock mathematical secrets. Grasping this theorem will significantly enhance your understanding of modular arithmetic and cryptography. Remember, practice makes perfect; the more you work with examples, the clearer the concepts become.

4 Diophantine Equations

Imagine trying to solve a puzzle where the pieces are integers, and the solution must be whole numbers only. That's essentially what Diophantine equations are all about. They're named after Diophantus of Alexandria, a mathematician who lived around the third century AD, and they've fascinated mathematicians ever since. These equations demand integer solutions; no fractions, no decimals, just whole numbers. Finding these solutions can be surprisingly tricky, even for seemingly simple equations.

The simplest type is the linear Diophantine equation, which takes the form $ax + by = c$, where a, b, and c are integers. The core question is: Do integer solutions for x and y exist? If they do, how many are there? Let's consider a concrete example: $3x + 5y = 1$. You can try different integer values for x and y, and you'll eventually find that $x = 2$ and $y = -1$ works ($32 + 5(-1) = 1$). But are there others? It turns out there are infinitely many! A key idea here is the concept of the greatest common divisor (GCD) of 'a' and 'b'. If 'c' is a multiple of the GCD of 'a' and 'b', then solutions exist. Otherwise, there are no integer solutions at all. Once you find one solution, you can generate infinitely more using a simple formula involving the GCD. This allows us to systematically explore all possible solutions.

Moving beyond linear equations, things get considerably more complex. Consider the Pythagorean triples, which are integer solutions to the equation $x^2 + y^2 = z^2$. This equation, representing the Pythagorean theorem, has a rich history and surprising connections to other areas of mathematics. Finding all Pythagorean triples involves a deep dive into the properties of integers and their factorization. You can use a parametrization method, using two integer parameters to generate all possible triples. Each of these parameters contributes in its own way to the final triple; one essentially represents the legs and another the hypotenuse. It's a fascinating way to generate all Pythagorean triples. These are just a few examples of Pythagorean triples; and it turns out that there are infinitely many such triples.

Nonlinear Diophantine equations are far more challenging. Many such equations have no known general solutions and are areas of ongoing mathematical research. Take Fermat's Last Theorem, a famously difficult problem that remained unsolved for centuries. It states that there are no integer solutions to the equation $x + y = z$ for any integer value of n greater than 2. This seemingly simple statement was finally proven by Andrew Wiles in the 1990s, demonstrating the incredible difficulty that can be hidden within seemingly simple equations. The study of Diophantine equations bridges various areas of number theory. The search for solutions often involves prime factorization, modular arithmetic, and the properties of integers. For example, understanding the divisibility of numbers can significantly affect the existence and nature of solutions. The techniques used to solve these equations, such as the Euclidean algorithm, which is central to finding the greatest common divisor, also have widespread use across multiple mathematical fields. Furthermore, Diophantine equations provide a fascinating playground to explore the elegance and complexity of number theory, which opens up avenues of exploration in a range of contexts.

Moreover, Diophantine equations have surprising applications beyond pure mathematics. They have appeared in coding theory and cryptography, playing crucial roles in maintaining data security. Because of the way their solutions are structured and because solutions can be so hard to compute, they provide a powerful tool for devising algorithms for encryption and decryption. The difficulty of finding integer solutions for certain types of equations forms the basis of many modern encryption systems, providing security for financial transactions, sensitive data, and other crucial information. Exploring these connections is a great way to appreciate the power and relevance of abstract mathematical concepts.

In summary, Diophantine equations represent a captivating blend of challenge and elegance, offering beginners a glimpse into the rich and intricate world of number theory. From the relative simplicity of linear equations to the profound depths of nonlinear ones, these equations provide a compelling illustration of the beautiful complexity of integers, underlining their importance in mathematics and beyond. The journey to solving them provides practical experience in various number theory techniques, highlighting the underlying interconnectedness of seemingly disparate mathematical concepts. Understanding Diophantine equations provides not just a mathematical skill, but a deeper appreciation for the beauty and power of mathematical reasoning. Through solving these equations, you learn to appreciate the intricate tapestry that underpins the world of numbers.

4.1 Linear Diophantine Equations

Imagine you're trying to solve a puzzle where you only have whole numbers to work with. That's essentially what a Diophantine equation is - an algebraic equation where the solutions must be integers. This chapter focuses on the simplest kind, linear Diophantine equations, which take the form $ax + by = c$, where a, b, and c are integers. Finding solutions means discovering integer values for x and y that make the equation true. This might seem simple, but finding these solutions can be surprisingly tricky! For instance, consider the equation $3x + 5y = 1$. You can't simply guess and check your way to a solution. You need a system.

The key to solving these equations lies in the greatest common divisor (GCD) of a and b. Remember that chapter on divisibility? It's time to put that knowledge to the test. If the GCD of a and b doesn't divide c, then there are no integer solutions at all. Think of it like this: if you're trying to divide a pie into three pieces and then combine it with another that's been divided into five, there's no way to get exactly one slice using only whole pieces unless you're willing to use fractions. This principle is fundamental to understanding solvable Diophantine equations.

But what if the GCD does divide c? Then, we can find a solution using the Euclidean algorithm, a method you might recall from earlier chapters. It's a process that helps us find the GCD by repeatedly applying the division algorithm. It might seem like extra work, but using the Euclidean algorithm provides a solid mathematical method of finding a solution that guarantees correctness and avoids the unpredictable guessing approach. It's a powerful tool, particularly when dealing with larger numbers where guesswork would become impractical. Once you have one solution, you can find infinitely many others. This is because adding or subtracting multiples of b/GCD(a,b) to x, and the opposite to y (subtracting multiples of a/GCD(a,b) from y) will still satisfy the equation.

Let's tackle an example. Let's say we have the equation $6x + 9y = 15$. First, we find the GCD(6, 9) which is 3. Since 3 divides 15, solutions exist. Now use the Euclidean algorithm (or simply notice) to find that $6(2) + 9(-1) = 3$. Multiplying this by 5, we get $6(10) + 9(-5) = 15$. Therefore, $x = 10$ and $y = -5$ is a particular solution. Now, since the GCD is 3, any solution can be expressed as $x = 10 + 3k$ and $y = -5$ \item 2k, where k is

an integer. If you plug different values of k into this equation, you'll see how it generates many other solutions.
So, to summarize the process: first, calculate the GCD of a and b. If it doesn't divide c, you're done; no integer solutions exist. If it does divide c, then the Euclidean algorithm helps in finding one particular solution. From this single solution, an infinite number of other solutions can be generated.
However, understanding linear Diophantine equations is just a stepping stone in number theory. While they might not seem to hold the same profound implications as some of the topics that lie ahead, understanding them thoroughly lays a strong foundation for tackling more complicated equations in later chapters. They are essential for gaining insight into the intricacies of number theory and provide a crucial link to more complex ideas that will surface later. The skills and techniques learned here will prove to be invaluable as we delve into the world of more challenging problems within number theory. Their deceptively simple appearance hides a wealth of fundamental concepts and solutions. This lays a vital framework for your exploration of number theory's fascinating complexities.

4.2 Pythagorean Triples

A Pythagorean triple is a set of three positive integers (a, b, c) that satisfy the equation $a^2 + b^2 = c^2$. Think of it like this: you have a right-angled triangle, and these numbers represent the lengths of its sides, with c being the longest side, the hypotenuse. Finding these triples is like a mathematical treasure hunt, uncovering integer solutions to a fundamental geometric relationship. Euclid, that ancient Greek mastermind, figured out a way to generate these triples, essentially providing a formula that works every time.
His method involves using two positive integers, m and n, where m is greater than n. You then calculate a = m^2 \item n^2,

$b = 2mn$, and $c = m^2 + n^2$. Any values you choose for m and n (provided m > n), will yield a Pythagorean triple. Let's try it: if $m = 2$ and $n = 1$, you get $a = 3$, $b = 4$, and $c = 5$, the most famous Pythagorean triple of them all. Notice how neatly this fits the Pythagorean theorem: $3^2 + 4^2 = 9 + 16 = 25 = 5^2$.

But Euclid's formula doesn't give all Pythagorean triples. There are others, hiding in plain sight, waiting to be discovered. Some triples, like (3, 4, 5) or (5, 12, 13), are relatively straightforward. Others are much larger and less obvious; finding them requires more sophisticated methods, perhaps a computer program to search for solutions. The beauty of it lies in the sheer infinity of potential solutions. There are infinitely many Pythagorean triples, each a unique combination of integers satisfying the fundamental equation. This unboundedness is what makes this area of number theory so endlessly fascinating.

This concept ties directly into the broader themes of Diophantine equations, mathematical puzzles that demand integer solutions. Pythagorean triples are the simplest example of this type of equation, a gateway to more complex problems. Understanding how to find and generate them provides a foundational understanding of how mathematicians approach such problems, laying the groundwork for tackling more advanced Diophantine equations, equations far less well understood. The process of systematically exploring possibilities, testing conditions, and identifying patterns, is central to solving such problems.

While simple in concept, the implications of Pythagorean triples reach far beyond basic geometry. They have unexpected applications in various fields. For instance, in computer graphics, understanding Pythagorean triples is crucial for efficiently calculating distances and positions within a 2D or 3D space. This allows for smooth animation, accurate rendering, and efficient simulations. The ability to use integers to represent geometric relationships simplifies calculations and minimizes the need for complex, potentially error-prone floating-point operations. The efficient computation of Pythagorean triples, therefore, has practical implications for fields that rely heavily on computational geometry.

Beyond computer graphics, Pythagorean triples are used in cryptography, the art of secure communication. Some cryptographic algorithms leverage properties of these triples to secure data. Because of the unique characteristics of these triples, they contribute to more secure and reliable methods of data encryption and decryption. Their inherent mathematical properties ensure higher levels of security, protecting sensitive information in ways traditional methods might not be able to.

The search for and analysis of Pythagorean triples exemplifies the very essence of mathematical exploration: starting with a seemingly simple concept, unearthing deeper connections, and discovering practical applications. It teaches patience, perseverance and attention to detail, qualities crucial to tackling any mathematical or real-world problem. The journey of exploring these triples is not just about the destination, finding the triples themselves, but the intellectual journey of discovery.

The ability to generate triples and understanding their properties is a key to unlocking further complexities in number theory, highlighting the interconnectedness of seemingly disparate mathematical concepts. The exploration begins with the seemingly simple, but the reward is a profound appreciation for the intricate beauty and surprising power within the realm of numbers. The study of number theory offers a profound understanding that lies at the heart of mathematical reasoning.

5 Rational Numbers and Continued Fractions

Rational numbers, those familiar quotients of integers, possess a hidden elegance when expressed as continued fractions. Imagine a fraction like 17/5. We can express it not just as a simple fraction, but as a chain of simpler fractions nested within each other. This is the essence of a continued fraction representation. The process begins with division: 17 divided by 5 gives 3 with a remainder of 2. This gives us the first term, 3. Now, we invert the remainder's fraction (2/5) to get 5/2, and repeat the process. 5 divided by 2 is 2 with a remainder of 1. We obtain a new term, 2. We continue this process, inverting the remainder's fraction each time until the remainder is zero. For 17/5, this yields [3; 2, 2], meaning $3 + 1/(2 + 1/2)$. This seemingly intricate representation offers remarkable insights. Converting a rational number into a continued fraction is straightforward, employing a repetitive algorithm. However, the converse, transforming a continued fraction back into a rational number, requires careful attention to the order of operations. We can proceed from the end of the continued fraction expression, working inwards step by step, performing the indicated additions and divisions. The expression [3; 2, 2] leads to $3 + 1/(2 + 1/2) = 3 + 1/(5/2) = 3 + 2/5 = 17/5$. This confirms the accuracy of our method and reinforces your understanding of this process. Practice with different rational numbers will increase your fluency with this transformation process. Remember, accuracy is key; a single mistake can cascade through the calculations, leading to an incorrect final result.

Continued fractions provide incredibly accurate approximations of rational numbers. Consider the continued fraction representation of a number. By truncating this representation at various points, we generate a sequence of approximations known as convergents. These convergents provide progressively better approximations of the original rational number. For example, the convergents of [3; 2, 2] are 3, 3 + 1/2 = 7/2, and 17/5. You can clearly observe how each convergent progressively improves the approximation, converging towards the original rational number. This property is incredibly useful in situations where high precision is required, such as numerical computations or estimations. The error of approximation gradually reduces as you include more and more terms in the truncated expression.

The applications of continued fractions extend far beyond the realm of pure mathematics. They find practical uses in various fields. In computer science, for example, continued fractions simplify the process of representing rational numbers and operations on them, often leading to faster computations and greater efficiency. In cryptography, where secure communication hinges on the efficiency and accuracy of computations, the properties of continued fractions make them indispensable. In some approximation algorithms, continued fractions provide excellent rational approximations for irrational numbers, offering a level of precision that can be adjusted according to the needs of the specific application. They can also be instrumental in solving certain types of Diophantine equations, showcasing their ability to bridge the gap between pure theory and real-world applications.

Beyond their practical applications, continued fractions highlight the richness and interconnectedness within mathematics. They provide an alternative representation for rational numbers, demonstrating a different perspective on numbers we often take for granted. Understanding continued fractions reveals hidden structures and relationships, deepening your understanding of the fundamental nature of rational numbers, which forms an important basis for understanding more complex number systems. The elegance of this seemingly simple concept underscores the beauty and power of mathematical exploration and expands your horizons in the mathematical world. Mastering continued fractions not only enriches your mathematical toolset but also deepens your appreciation for the intricate beauty hidden within seemingly simple mathematical concepts. This opens up opportunities to further investigate advanced mathematical fields that build upon these foundations.

5.1 Continued Fraction Representation

Imagine a pizza cut into infinitely many slices. That's a bit like a continued fraction. It's a way of representing a number, especially rational numbers, as a sum of integers and fractions, where the denominators themselves are fractions, and so on, creating a seemingly endless nested structure. We start with a rational number, say 17/5. Divide 17 by 5: you get 3 with a remainder of 2. This 3 becomes the first integer in our continued fraction. Now, instead of dividing 2 by 5 directly, invert the fraction. That's 5/2. This repeats the process: 5 divided by 2 gives us 2 with a remainder of 1. So 2 is our next integer. We continue this process, inverting the remainder until it becomes 0. For this example, we get: $17/5 = 3 + 1/(2 + 1/2)$. This expression is the continued fraction representation. It's a beautiful and surprising way to represent something seemingly straightforward!
Converting a rational number to a continued fraction isn't

just a mathematical curiosity; it offers a practical pathway to approximate irrational numbers. For instance, consider the golden ratio (approximately 1.618). Its continued fraction representation is strikingly simple: 1 + 1/(1 + 1/(1 + ...)). Notice the elegant repetition of the pattern. This infinite repetition is a hallmark of some irrational numbers. This specific continued fraction provides increasingly accurate approximations by truncating it at various points. The first few approximations of the golden ratio, obtained from truncating this continued fraction, are: 1, 2, 1.5, 1.666..., 1.6, and so on. Observe how quickly the approximations converge to the actual value. These approximations, called convergents, are rational numbers incredibly close to the irrational number, which is useful in various applications.

Now let's consider the opposite: turning a continued fraction back into a standard rational number. This is achieved systematically by working our way back from the inside, simplifying the fraction step-by-step. For example, if we're given the continued fraction [1;2,3,4], which means 1 + 1/(2 + 1/(3 + 1/4)), we first calculate 3 + 1/4 = 13/4. Then we compute 2 + 1/(13/4) = 2 + 4/13 = 30/13. Finally, 1 + 1/(30/13) = 1 + 13/30 = 43/30. Therefore, [1;2,3,4] equals 43/30. This systematic approach lets us convert the unusual form into a well-known and easily understood rational number. This is fundamental to utilizing continued fractions in practical applications.

The power of continued fractions extends beyond simple rational numbers. They play a crucial role in approximating irrational numbers, providing a series of rational convergents that get increasingly closer to the target irrational value. Think of it as building a staircase that gradually approaches a specific, unreachable point. Each step, a rational convergent, gets you closer. This makes them invaluable in situations where precise calculations involving irrational numbers may be impractical, especially in applied fields where approximations are satisfactory. Continued fractions offer a way to manage those challenges elegantly. This process is efficient and provides practical approximations which are extremely useful in various areas.

Their applications are diverse. In computer graphics, continued fractions aid in efficient line drawing algorithms. They help in determining the optimal slope for lines that need to be displayed using pixel coordinates, a discrete and not continuously variable value. They are also helpful in digital signal processing. Continued fractions help analyze and process signals by allowing efficient representations of rational transfer functions commonly used in filter designs, making computations simpler and faster. The precision of continued fraction approximations means that digital signal processing filters can be designed with higher accuracy and efficiency. These are just a few of their applications. Continued fractions offer a unique and surprisingly practical tool in a world often governed by approximations. Their elegance and power highlight the rich interplay between theoretical mathematics and real-world problem-solving. Their potential to be used widely is continuously being explored. The versatility and deep usefulness are quite impressive.

5.2 Convergents and Approximations

Understanding continued fractions is key to mastering approximations. Imagine a fraction, say 22/7, a common approximation for . But how do we find better approximations? Continued fractions provide a systematic way. They represent a number as a sum of fractions, each with a numerator of 1, allowing us to progressively refine our approximation. Each stage in this process yields a convergent, a fraction increasingly closer to the true value.

Let's illustrate with an example. We can represent the golden ratio, (approximately 1.618), as a continued fraction: 1 + 1/(1 +

1/(1 + 1/(1. . .))). Notice the repeating pattern. The first few convergents are 1, 2, 3/2, 5/3, 8/5... Observe how each successive convergent gets closer to the actual value of . This method offers a powerful tool for approximating irrational numbers, offering increasingly accurate results with each step.
The beauty lies in the simplicity and efficiency. Calculating convergents is straightforward, requiring only basic arithmetic operations. This contrasts sharply with other approximation techniques that might involve complex calculations or iterative processes. Moreover, continued fractions provide not just an approximation but a sequence of increasingly better ones, giving you control over the desired accuracy. This controlled approach is invaluable in practical applications.
Consider the problem of approximating the square root of 2. While its decimal representation is infinite and non-repeating, continued fractions offer a concise representation and a method for calculating arbitrarily close approximations. The continued fraction for 2 is surprisingly simple: 1 + 1/(2 + 1/(2 + 1/(2. . .))). The convergents are 1, 3/2, 7/5, 17/12, 41/29. . . which progressively converge towards 2, providing ever more precise rational approximations.
The process is not limited to irrational numbers; it works equally well for rational numbers. In fact, every rational number has a finite continued fraction representation. This allows us to understand the relationship between rational numbers and their continued fraction forms, further solidifying their theoretical importance and practical uses. The elegance and efficiency of the method make it a fundamental tool in number theory and beyond.
Beyond simple approximations, convergents reveal deeper insights into the nature of numbers. They provide a way to analyze the structure of irrational numbers, showcasing their inherent complexities in a remarkably straightforward manner. The sequence of convergents allows us to study the rate of convergence, providing valuable information about the spe-

cific irrational number being approximated.

The practical applications are numerous. In computer science, continued fractions are used in algorithms for efficient rational approximations of real numbers. This is particularly useful in situations where exact values are impractical or impossible to compute, enabling a balance between accuracy and computational efficiency. In engineering, continued fractions can be employed in designing filters, control systems, and other applications requiring precise numerical approximations.

Continued fractions have even found their way into cryptography, where their properties have been exploited for certain cryptographic protocols and algorithms. The inherent structure and properties of the convergents allow for the generation of sequences with specific characteristics, potentially useful in creating secure and robust systems.

Furthermore, continued fractions have been used in approximation theory, providing rational approximations for various types of functions and providing effective methods for numerical analysis. They are applied in many areas, such as the design of approximation algorithms for complex mathematical functions, solving differential equations, and solving numerical integration problems.

The power of convergents lies not only in their practical ability to generate approximations but also in the structural insight they provide. They connect the seemingly disparate worlds of rational and irrational numbers, allowing us to explore the fine line between these two fundamental types of numbers, revealing hidden connections that may not be apparent through traditional approaches. The ability to systematically generate increasingly accurate approximations holds enormous value across diverse disciplines. It's this versatility and underlying elegance that elevates continued fractions from a mere approximation technique to a fundamental tool in mathematics.

5.3 Applications

Continued fractions, seemingly abstract mathematical objects, possess surprising practical applications. Consider their use in approximating irrational numbers. For instance, the famous constant , an irrational number, can be approximated with incredible accuracy using continued fractions. Each convergent, a rational number derived from the continued fraction expansion, provides a progressively better approximation. This is invaluable in fields like computer graphics, where precise representations of circular shapes and other curved figures depend heavily on accurate approximations of . Imagine the limitations if we only had crude estimates; the smoothness of those curves would suffer greatly.

The elegance of continued fractions extends to their use in solving Diophantine equations. Remember those equations from the previous chapter, where we seek integer solutions? Continued fractions offer a systematic approach to finding solutions, or determining whether solutions even exist. This is particularly useful in optimization problems, where constraints often necessitate integer solutions. For example, consider the distribution of resources - allocating a certain number of items among different groups, where the number of items and groups are fixed. Continued fractions can help find integer solutions that satisfy specific conditions, leading to the most efficient distribution. The precision here is not just mathematical; it has direct real-world impacts in areas like logistics and resource management.

Furthermore, the structure of continued fractions reveals deep insights into the properties of rational numbers. Analyzing the sequence of partial quotients (those numbers in the continued fraction representation) gives us unique information about the original rational number. This is not just theoretical; these patterns can be exploited in cryptography. For example, specific patterns in the continued fractions might be used to generate cryptographic keys or to analyze the strength of existing ones. The security of communications systems, the confidentiality of data transactions—these depend on the complex interplay of numbers, and continued fractions offer a unique and powerful lens through which to view this interplay. The development of new cryptographic techniques often requires exploring such hidden connections.

Beyond cryptography, continued fractions find application in the study of musical intervals. Musical harmony is, at its core, the precise mathematical ratio of frequencies. These ratios can be expressed and analyzed using continued fractions, thus helping to understand and generate musical scales and harmonies. Think of composing music; tuning instruments requires precision. Continued fractions provide a framework for constructing the most harmonious and pleasing melodies. This intersection of mathematics and music might seem surprising to some, but the underlying structure reveals a profound connection between seemingly disparate fields. This demonstrates how abstract mathematical concepts have unexpected and tangible real-world impacts. The ability to translate theoretical knowledge into functional tools highlights the importance of understanding abstract principles.

Continued fractions also surface in fields as diverse as astronomy and computer science. In astronomy, calculating the precise positions of celestial bodies sometimes requires high-precision approximations of irrational numbers, where continued fractions again prove invaluable. Similarly, in computer science, continued fractions play a role in the development of efficient algorithms for tasks involving numerical approximations. This broad applicability underscores the fundamental nature of continued fractions within mathematics and its pervasive impact across scientific disciplines. Their use in these diverse fields isn't merely coincidental; it highlights the underlying mathematical structure that governs many natural and engineered phenomena. The elegance and usefulness of this seemingly simple representation should not be underestimated.

The power of continued fractions lies not just in their ability to approximate irrational numbers with precision, but also in their inherent elegance and efficiency. They provide a concise and insightful way to represent rational numbers, facilitating a deeper understanding of their properties. This efficiency translates directly into practical applications, where speed and accuracy are often crucial. Whether we're building secure communications systems, composing music, or modeling celestial movements, the elegance and utility of continued fractions shine through. Their applications are a testament to the unexpected and profound connections that exist within the tapestry of mathematics. The exploration of these connections continues to yield exciting new developments, continually enriching our understanding of the world around us.

6 Introduction to Modular Forms

Imagine a perfectly symmetrical tile, endlessly repeated to cover a flat surface. Modular forms are similar; they're complex functions exhibiting remarkable symmetry under transformations of the complex plane. Understanding them is key to unlocking deeper secrets within number theory. This chapter offers a gentle introduction, suitable for beginners. We won't dive into rigorous proofs, but focus on developing intuitive understanding.

Modular forms live in a world defined by modular groups. These groups, essentially collections of transformations, act like mirrors reflecting the form back onto itself. The most fundamental modular group, denoted SL(), consists of 2x2 matrices with integer entries and determinant 1. These matrices represent transformations on the complex upper half-plane; multiplying a complex number by such a matrix yields a new complex number, and remarkably, the modular form's value at these transformed points behaves predictably. It's a captivating dance of symmetry.

Think of Eisenstein series. These are specific examples of modular forms; their definitions involve infinite sums of terms, each related to the transformation of the complex plane. These sums, though appearing complicated, exhibit beautiful, predictable behavior under the actions of the modular group. You might visualize this as a kaleidoscope, each turn revealing a new, yet harmoniously related, pattern. This predictable behavior is a defining characteristic of modular forms themselves. They're not just any functions; they possess a profound elegance and order.

What's fascinating is the relationship between modular forms and L-functions. L-functions, often complex, are associated with many objects in number theory, from prime numbers to elliptic curves. The connection? A modular form can give birth to an L-function, and properties of the modular form directly influence the properties of the associated L-function. It's a deep link, linking the world of complex analysis (the realm of modular forms) to the world of arithmetic (the realm of L-functions and number theory's core problems).

The L-function associated with a modular form encodes deep arithmetic information. For example, consider Riemann's Hypothesis, one of the most important unsolved problems in mathematics. This hypothesis focuses on the distribution of the zeros of the Riemann zeta function, a specific type of L-function. While we are far from proving this hypothesis, the connection between modular forms and L-functions provides powerful tools and new perspectives to approach the challenge. It's like having a new map to navigate a previously uncharted territory.

To begin grasping modular forms, start by visualizing their symmetries. Think about how they remain unchanged, or transform in predictable ways, under specific transformations. Focus on understanding the basic modular group SL() and its action on the complex plane. Picture how the fundamental Eisenstein series exhibit that symmetry and recall their definition—an infinite sum—and try to see the pattern within.

Then, contemplate the significance of the link between modular forms and L-functions. This link is a powerful bridge, connecting complex analysis and the very heart of number theory. The profound implications of this connection highlight the unifying power of mathematics, revealing hidden relationships between seemingly disparate concepts. By studying these connections, we gain a deeper appreciation for the intricate beauty and the underlying harmony within mathematics.

Ultimately, modular forms are far more than just mathematical objects; they represent a deep interconnection between diverse branches of mathematics. Their study is a journey into the heart of number theory, revealing profound symmetries and connections that enrich our understanding of the world of numbers, and more. It's a journey filled with elegance, symmetry, and the surprising beauty of mathematical structures. The more you explore them, the more profound their implications become, opening doors to a deeper understanding of number theory's intricate tapestry.

6.1 Modular Groups

Modular groups form the backbone of understanding modular forms, those elegant mathematical objects with surprising connections to various fields. Think of them as the stage upon which the drama of modular forms unfolds. They are groups of transformations, specifically linear fractional transformations, acting on the complex upper half-plane. These transformations are like rotations and translations, but far more intricate. Imagine stretching and warping the complex plane, twisting and turning it, yet leaving certain properties, like the fundamental structure of the modular form, unchanged. We represent these transformations using matrices; each matrix corresponds to a specific warping action. The group structure means we can combine these transformations - sequentially applying one after another - and the result is another valid transformation, still within the modular group. This group structure, this ability to combine transformations, is crucial to understanding the symmetries inherent in modular forms. This systematic study allows us to categorize and analyze modular forms more effectively.

The simplest example of a modular group is the modular group SL(2,). This group comprises all 2x2 matrices with integer entries and a determinant of 1. Each of these matrices represents a linear fractional transformation. These transformations act on complex numbers; a complex number 'z' transforms into (az+b)/(cz+d), where a, b, c, and d are the entries of the matrix. This transformation might seem complex, but it represents a fundamental warping operation, twisting and reshaping the upper half of the complex plane. The restriction to matrices with a determinant of 1 ensures that these transformations preserve the fundamental structure and properties of the complex plane. The group operation is matrix multiplication—the combination of two transformations is simply the product of their corresponding matrices. Notice the elegance; the complex geometry of the transformations is neatly encoded in simple matrix algebra.

Understanding how these matrices act on the complex upper half-plane is key. The transformations aren't just arbitrary mappings; they possess a remarkable property: they preserve the shape of certain regions within the upper half-plane. This preservation is directly related to the symmetries of modular forms, and understanding this connection is a cornerstone of the theory. This is where the "modular" aspect comes into play. These transformations, by their very nature, map certain fundamental regions onto themselves, revealing inherent symmetries in the structure of the upper half-plane and its connections with modular forms. This invariance under modular transformations is what truly defines a modular form; these are functions that remain unchanged, or at most undergo a simple transformation (like multiplication by a factor), after the application of these transformations.

Let's delve deeper into the structure of SL(2,). This group, while seemingly simple in definition, is surprisingly rich and complex. It's not just a collection of matrices; it's a group with a fascinating internal structure. It possesses generators, meaning that every single element of the group can be expressed as a finite combination of just two basic transformations. Identifying these generators simplifies the analysis of the entire group considerably. This simplifies the seemingly infinite set of transformations into a manageable structure. It's like understanding how a complex machine works by understanding its basic parts and how they interact. This approach drastically reduces the complexity of working with an infinite group and opens avenues for deeper analysis.

Consider the two matrices: one representing a translation and the other a rotation or inversion. These seemingly simple transformations, when combined repeatedly, generate every possible element of the modular group SL(2,). This fact allows us to approach the infinite structure of the group in a systematic and manageable manner. Moreover, understanding the generators is crucial for classifying and studying the properties of modular forms associated with these groups. Knowing the generators allows mathematicians to work more efficiently and to gain a deeper understanding of the properties of modular forms. This efficient organization lets us analyze modular forms with greater precision and efficiency.

Beyond SL(2,), there are other modular groups. These groups are often subgroups of SL(2,), inheriting its structure but with additional restrictions. For instance, we might consider groups of matrices with larger determinants, or those whose entries satisfy specific congruences. Each group creates a different set of transformations, leading to various types of modular forms. These different modular groups and subgroups give rise to a rich and diverse landscape of modular forms, each with its own unique properties and applications. It is this vast array of structures that unlocks the power and versatility of

modular forms within number theory and other mathematical disciplines. The study of these subgroups reveals a hierarchy of modular structures, which adds further depth to the theory. Each subgroup carries its own intricate patterns, offering mathematicians new challenges and deeper insights.
The study of modular groups is not merely abstract algebra; it's a journey into the heart of number theory. These groups provide a framework for understanding the symmetries and structures that underpin modular forms and, by extension, various seemingly unrelated mathematical and even physical phenomena. The profound connections between seemingly disparate fields, revealed through the lens of modular groups, underscore the deep interconnectivity of mathematics. It highlights the elegance and power of mathematics to unify seemingly distinct concepts and opens the door to even more exciting discoveries and applications. The journey into the depths of modular groups is not just an exploration of abstract mathematics; it is an exploration of the deep connections and elegant structures underlying many mathematical fields, offering a glimpse into the beauty and harmony of the mathematical universe.

6.2 Eisenstein Series

Imagine a perfectly symmetrical tile pattern, endlessly repeating across an infinite plane. That's the kind of elegant structure we find in modular forms, and Eisenstein series are some of the most basic, yet profoundly important, examples. They're special functions, built from sums that seem simple at first glance, but hold deep mathematical secrets. Think of them as the foundational building blocks for a vast and complex structure within number theory.
These series are defined using a rather straightforward formula, involving sums over integers. You'll start by considering

a variable, often denoted as 'z', which is a complex number—a number with both a real and an imaginary part. The formula itself involves summing terms of the form (1/mz + nz)^k, where 'm' and 'n' are integers, and 'k' is a positive integer called the weight. This simple-looking formula, when considered carefully, can generate remarkably intricate and powerful results. Mastering this formula is key to understanding a whole universe of number theory.

The weight 'k' plays a crucial role in determining the properties of the Eisenstein series. Different values of k lead to series with vastly different behaviors. For instance, when k is even and greater than 2, the Eisenstein series converge and become modular forms, exhibiting stunning symmetries. Explore different values of 'k'; you will discover fascinating patterns. Each value will unlock a new perspective and a deeper understanding of the functions. Experimenting and observing these patterns is incredibly rewarding.

The beauty of Eisenstein series lies in their inherent symmetry. They possess a property called modular invariance. This means their value remains unchanged under specific transformations of the complex variable 'z'. These transformations are related to the modular group, a fascinating mathematical object in itself. Understanding the modular group and its connection to Eisenstein series is a vital step in unlocking the deeper meaning of this concept. It unlocks a pathway to understanding the more complex aspects of this profound mathematical universe.

Calculating Eisenstein series can be surprisingly straightforward, especially for low values of 'k'. However, as 'k' increases, the complexity grows rapidly. There are sophisticated techniques for calculating these series, involving advanced concepts from complex analysis. But even calculating simpler instances reveals the underlying beauty and regularity of these mathematical objects. The process of calculation itself helps solidify your understanding of the concepts involved.

Eisenstein series aren't just abstract mathematical curiosities; they have tangible applications. They appear in various areas of mathematics and beyond. One crucial application is their connection to other modular forms. They provide a starting point for building more elaborate modular forms. Furthermore, Eisenstein series possess remarkable relationships with partitions and elliptic curves, a connection hinting at the profound interweaving of seemingly unrelated mathematical concepts. Therefore, mastering Eisenstein series is vital not only for a deeper understanding of modular forms, but also for comprehending their connections to other important mathematical objects.

These relationships aren't merely theoretical; they have practical implications in areas such as number theory, cryptography, and even some theoretical physics. However, that will be discussed in detail in later chapters of this book. For now, understanding the fundamentals of Eisenstein series, its symmetries and how they function is of paramount importance. The understanding you gain here, of Eisenstein series and modular forms, will serve as a bedrock upon which you can construct a much broader understanding of number theory and the interconnectedness of mathematics and its applications. This foundational knowledge is crucial for continued progress in the field.

So, as you delve deeper into the world of modular forms, remember the Eisenstein series. They're not just a starting point; they're a powerful lens through which we can view the intricate and beautiful tapestry of numbers. They represent the essential stepping stones to mastering the rich, intricate, and surprisingly interconnected world of number theory. They are the first steps on your journey into the wonders of modular forms; the rest is yours to explore and discover. Embrace the challenge, and you'll be rewarded with a profound appreciation for the elegance and power of mathematics.

6.3 Modular Forms and L-functions

Imagine a beautiful, intricate tapestry, its threads interwoven to create a breathtaking design. That's number theory, and modular forms are some of its most vibrant threads. They're complex functions that possess a remarkable symmetry, transforming beautifully under the action of modular groups. These groups, essentially transformations of the complex upper half-plane, reveal hidden connections and structures within the seemingly chaotic world of numbers.

Understanding modular forms requires a grasp of complex analysis, but their core essence lies in their periodic nature. Just like how a repeating pattern on wallpaper seamlessly continues, modular forms exhibit a similar repetition, but in a more sophisticated setting of complex numbers. This periodicity isn't arbitrary; it's governed by the modular group, its actions shaping the forms' behavior. Their profound symmetries unveil deep patterns in number theory.

We've explored modular groups; now let's delve into Eisenstein series, a specific type of modular form. They are relatively easy to understand and serve as excellent examples to demonstrate the key properties of modular forms. These series are constructed by summing certain powers of the complex numbers, specifically focusing on the terms involving the action of the modular group. Their convergence properties, beautifully intertwined with complex analysis, directly relate to their modularity. You'll see how seemingly complex calculations neatly simplify due to the inherent symmetries of the Eisenstein series. Exploring these series provides a concrete example, grounding the abstract ideas of modular forms in a tangible mathematical object.

Now, the link between modular forms and L-functions forms the heart of this chapter. L-functions, deceptively simple in appearance, are complex functions related to the distribution of prime numbers and other arithmetic objects. These functions are not just a mathematical curiosity; they pack a powerful punch. They encode deep information about the underlying number-theoretic structures. Think of them as secret codes revealing the distribution of primes.

The relationship between modular forms and L-functions is profound, as proven by the landmark modularity theorem (previously Taniyama-Shimura conjecture). This theorem states that every elliptic curve, a type of geometric object described by certain polynomial equations, corresponds to a modular form. This seemingly simple statement has monumental consequences, including the spectacular proof of Fermat's Last Theorem, a problem that had baffled mathematicians for centuries. It's a shining example of how different areas of mathematics, seemingly disparate, can be intricately connected. The modularity theorem provided a powerful bridge linking seemingly unrelated concepts.

The power of this connection lies in the ability to translate problems from one domain to the other. A difficult problem concerning an elliptic curve can often be transformed into a tractable problem regarding its corresponding modular form, and vice-versa. This interoperability opens new avenues for research and helps unravel deep mysteries within number theory. This "translation" is a powerful tool in the mathematician's toolbox, allowing one to view a problem from a different, potentially more illuminating, perspective.

Let's consider a simple illustration. Imagine trying to understand the distribution of prime numbers. While the Prime Number Theorem provides an asymptotic estimate, precise results are harder to obtain. The connection between modular forms and L-functions allows us to approach this problem through the analysis of the corresponding modular forms, often offering refined and more accurate results. This approach opens new and exciting possibilities for understanding the distribution of primes, using modular forms as analytical tools.

The study of modular forms and their connection to L-functions continues to be a significant area of research. These objects are not merely mathematical curiosities. Their deep connections to elliptic curves, Galois representations, and other areas of mathematics highlight their profound significance. For instance, modular forms have found applications in cryptography. Their intricate symmetries provide a foundation for designing highly secure encryption schemes, protecting sensitive information in the digital age.

This chapter has provided just a glimpse into the fascinating world of modular forms and L-functions. This area remains an active field of research, with many open problems and ongoing investigations. Even though this chapter offered simplified explanations, you have now gained a solid foundation to appreciate the complexity and beauty of the connection between these powerful mathematical tools. This knowledge equips you to navigate further explorations into advanced number theory. The intricacies of modular forms await those who dare to explore. The journey into their depths will richly reward the curious mind. Their elegance is truly something to behold.

7 Applications and Further Explorations

The elegance of number theory extends far beyond the abstract realm of mathematical proofs. Its principles find powerful and unexpected applications in diverse fields, shaping our modern world in profound ways. Consider cryptography, for instance. The security of online transactions, confidential communications, and digital identities hinges on sophisticated algorithms rooted in number theory's core concepts. Prime numbers, in particular, play a crucial role; their unique factorization property underpins the strength of widely-used encryption methods like RSA, ensuring the privacy of sensitive information exchanged daily across the globe. The difficulty of factoring large numbers into their prime components forms the bedrock of this security. This computational challenge, a direct consequence of the inherent structure of prime numbers, makes breaking these codes practically infeasible, safeguarding our digital interactions. The seemingly abstract world of integers thus becomes a vital component of our secure digital lives.

Moving beyond encryption, consider the crucial role of number theory in coding theory. This field focuses on efficient and reliable data transmission, a necessity in our digital age. Error-correcting codes, essential for ensuring accurate data transfer over noisy channels, are often designed using algebraic structures directly related to number theory. Finite fields, for example, form the foundation for many powerful error-correcting codes such as Reed-Solomon codes. These codes are used extensively in various applications, from satellite communications where signal interference is a major concern, to compact disc technology where the preservation of digital audio is critical. This application demonstrates the practical importance of mathematical structures - seemingly abstract algebraic objects - in solving real-world problems, underlining the impact of number theory's elegant principles on everyday technology.

Finally, the reach of number theory stretches even into the seemingly unrelated world of physics. While perhaps less immediately apparent, its connections are profound and increasingly significant. The study of prime numbers and their distribution, for example, has surprising parallels with the study of energy levels in chaotic quantum systems. Certain statistical properties observed in the distribution of prime numbers are found to mirror the behavior of these quantum systems, hinting at deep underlying connections between seemingly disparate areas of science. Furthermore, concepts from modular forms, complex mathematical objects with rich symmetry properties, have made unexpected appearances in string theory, a cutting-edge theoretical framework attempting to unify all fundamental forces of nature. While the exact nature of these connections is still actively being investigated, they clearly illustrate that number theory's influence transcends the boundaries of pure mathematics, providing valuable tools and insights into the fundamental laws governing our universe. The seemingly simple world of integers thus reveals its complexity and importance through these remarkable connections, extending its reach beyond pure mathematics and into the fabric of scientific understanding. Further investigation into these interdisciplinary relationships promises exciting new discoveries, unveiling more of number theory's unseen connections and demonstrating the power of its inherent elegance. These applications are just the tip of the iceberg. As we continue to explore the depths of number theory, we are sure to uncover even more astonishing and impactful connections to the world around us. The journey into the intricate tapestry of integers is far from over; it is an ongoing exploration, fueled by curiosity and promising countless future discoveries. The journey of exploration in the world of number theory continues and holds further potential. The further study of these relationships promises an enriching and revealing experience. The implications are vast and hold potential.

7.1 Cryptography

Cryptography relies heavily on number theory, specifically modular arithmetic and prime numbers. Imagine trying to send a secret message; you wouldn't want just anyone to read it. Cryptography provides the tools to protect your information. One fundamental technique uses prime numbers to create keys. These keys, incredibly large prime numbers, are practically impossible to factor into their constituent primes within a reasonable timeframe. This difficulty forms the bedrock of many modern encryption methods.
The RSA algorithm, a widely used public-key cryptosystem, is a prime example. It leverages the properties of modular arithmetic and the difficulty of factoring large numbers. To encrypt a message, you use a public key, which is freely available. Decryption, however, requires a private key, a closely guarded secret derived from the same prime numbers that created the public key, but not directly derivable from it. The security lies in the computational infeasibility of factoring the product of two extremely large primes, even with today's most advanced computers. This asymmetry between encryption and

decryption is crucial for secure communication.
Consider a simpler scenario. Let's say you want to encode a single digit message. You could use a simple substitution cipher, where each digit maps to another digit according to a secret rule. You could establish a key that transforms each digit. For example, 0 becomes 7, 1 becomes 3, 2 becomes 9 and so on. This is easy to implement, but relatively easy to break. A more sophisticated technique involves modular arithmetic. Suppose your key is a prime number, let's say 11. You could encrypt a digit x by calculating x modulo 11. This wraps the result around 11, providing a level of obfuscation. To decrypt, you need to find the inverse of the modular operation, again depending on the properties of the prime number.
The beauty of prime numbers in cryptography isn't just about their sheer size, making factoring difficult. It's about their mathematical properties, such as their unique role in factorization, which makes them ideal for creating one-way functions. A one-way function is easy to compute but extremely hard to reverse without special information, like the private key in RSA. This is precisely why prime numbers form such a crucial part of the foundation of modern cryptography.
Beyond RSA, many other cryptographic techniques employ number-theoretic concepts. Elliptic curve cryptography, for instance, utilizes the properties of elliptic curves defined over finite fields. These curves possess intricate mathematical structures that create strong cryptographic systems, often requiring less computational power than RSA for similar levels of security. The elegance of using these geometric objects in cryptography speaks volumes to the unexpected power of number theory.
Understanding the mathematical principles underlying these algorithms is not just about memorizing formulas; it is about grasping the fundamental concepts. You are not merely manipulating equations but manipulating the very fabric of numerical structures. The core idea behind cryptographic security

is computational complexity: making the decryption process computationally intractable for unauthorized access, ensuring the secrecy of the information.

The field of cryptography is constantly evolving, with new algorithms and techniques being developed to counter emerging threats and enhance security. However, at the heart of it all remains the elegant and powerful mathematics of number theory. Mastering the fundamentals gives you a deep appreciation for this essential area of security, providing you with a strong foundation for further explorations into advanced cryptographic methods. Learning about these fundamentals and seeing their surprising applications helps us understand the deep connections between seemingly abstract mathematical concepts and their vital role in securing our digital world. The journey through number theory opens doors to unexpected areas, emphasizing the profound utility of theoretical mathematics in everyday life. It is a journey well worth undertaking.

7.2 Coding Theory

Imagine sending a message across a noisy channel. Maybe it's a radio signal, a computer network, or even a whispered conversation in a crowded room. The message, initially perfect, arrives distorted, with bits flipped or missing. Coding theory is the mathematical framework that allows us to not only detect these errors but also correct them, ensuring the message's integrity. It's a powerful tool born from number theory's elegant structures, proving that seemingly abstract concepts hold incredible real-world relevance.

At its core, coding theory relies on cleverly constructed codes. These aren't secret codes like those in spy novels; instead, they're systematic ways of adding extra information to the original message, creating redundancy. This redundancy acts as a safeguard, allowing us to pinpoint and fix errors introduced during transmission. Think of it like sending multiple copies of the same message, each slightly altered. Even if some copies are corrupted, we can compare them, identify the inconsistencies, and reconstruct the original. This is the basic idea behind error-correcting codes.

One simple, yet effective, example is the repetition code. To send a single bit of information, say a '1', we transmit '111' instead. If one bit gets flipped during transmission—let's say we receive '011'—we can easily see that the majority of the received bits are '1', allowing us to correctly decode the message. While straightforward, repetition codes are inefficient for larger messages, as they significantly increase the amount of data sent. More sophisticated codes offer better efficiency by leveraging the properties of number fields, finite fields and modular arithmetic.

We can delve deeper into more advanced codes like Hamming codes. They're more efficient than simple repetition codes, detecting and correcting single-bit errors within a set of data. These are built on concepts of parity checks, creating a kind of mathematical checksum for our data. By strategically adding parity bits, Hamming codes allow us to pinpoint the location of a single error and correct it. This clever approach balances redundancy with efficiency, resulting in a robust system capable of handling noisy channels.

Consider the challenge of sending data to a distant spacecraft. The signal, weakened by vast distances and cosmic interference, is prone to errors. Sophisticated error-correcting codes, based on principles from coding theory and finite fields, are crucial for ensuring the integrity of the data received back on Earth. These codes, intricately designed, add just the right amount of redundancy to correct errors, while minimizing the extra data transmitted. The success of space exploration hinges on such mathematical elegance.

Another significant aspect of coding theory involves understanding the limits of error correction. The Shannon-Hartley theorem, a cornerstone of information theory, provides a fundamental limit on the amount of information we can reliably transmit over a noisy channel. It defines the channel capacity, the maximum rate at which information can be sent without error. This theorem establishes that even the most sophisticated code can't exceed this limit. It highlights the interplay between the code's design, the channel's noise level, and the achievable rate of transmission.

Moving beyond error correction, coding theory also addresses the problem of data compression. Source coding techniques aim to represent information using as few bits as possible, without losing essential data. Think of a JPEG image. It uses clever compression algorithms that exploit the redundancy within the image data to reduce the file size significantly. This ability to efficiently compress information is particularly important for transmitting large amounts of data, saving bandwidth and storage space. Efficient compression techniques minimize redundancy, reducing the overall data size and making transmission quicker.

Finally, let's consider the link between coding theory and cryptography. While seemingly distinct, they share a common thread: the careful manipulation of information to achieve a specific goal. Cryptography aims to protect information by making it unintelligible to unauthorized individuals. Error-correcting codes, on the other hand, aim to ensure the reliability of information transmission even in the presence of noise. Although distinct, these mathematical areas often intersect. Modern cryptography often employs sophisticated error-correcting codes to protect encrypted data during transmission. This connection demonstrates the surprising interconnectedness of seemingly different fields within the realm of mathematics. The application of these concepts showcases the power of fundamental mathematical principles in solving practical, real-world problems. Understanding coding theory opens doors to a fascinating world of communication, computation, and data security.

7.3 Physics

Unexpectedly, the elegant structures found within number theory find echoes in the fundamental laws governing our universe. It might seem like a vast leap - from the abstract world of integers to the concrete reality of physical phenomena - but the connections are surprisingly deep and profound. For instance, consider the distribution of prime numbers, a topic central to number theory. Their seemingly random scattering, governed by the Prime Number Theorem, bears an intriguing resemblance to the distribution of energy levels in complex quantum systems. While not perfectly analogous, the irregular spacing of primes mirrors the unpredictable nature of energy level spacing in chaotic systems, offering a fascinating point of comparison for physicists and mathematicians alike.

The concept of modular arithmetic, with its cyclical nature, provides another bridge. Think about the symmetries observed in crystals. These repeating patterns, governed by discrete rotational symmetries, can be elegantly described using modular arithmetic. The inherent cyclical nature of modular operations mirrors the periodic boundary conditions often employed in theoretical physics models, particularly in condensed matter physics when simulating crystal lattices. This makes modular arithmetic a surprisingly useful tool for analyzing complex systems.

Furthermore, the study of Diophantine equations, those with integer solutions, resonates with the discrete nature of many physical quantities. The quantization of energy, for example, means that energy levels in atoms aren't continuous but rather exist in distinct steps. This fundamental concept in quantum mechanics shares a parallel with the discrete solutions sought in Diophantine equations. Finding these integer solutions mirrors the process of determining permissible energy levels within a quantum system, showcasing the surprising interconnectedness of seemingly disparate fields. Consider the implications for the development of new materials or the understanding of quantum phenomena. The tools and techniques developed in the realm of Diophantine equations could potentially unlock new approaches to those seemingly intractable problems.

The connections extend even further. Continued fractions, a seemingly abstract mathematical tool, have found surprising applications in the analysis of dynamical systems, which deal with the behavior of systems that evolve over time, crucial in fields like astrophysics and fluid dynamics. The ability to express rational numbers using continued fractions and their convergents provides a powerful tool for approximating quantities within dynamical systems, aiding in predictions and analyses. The accuracy and efficiency of these approximations are directly relevant to predicting the trajectories of celestial bodies or modelling complex fluid flows. This speaks to the unifying power of mathematics in bridging diverse fields.

Finally, modular forms, with their rich mathematical structure and their connection to L-functions, offer an unexpected link to string theory, a highly theoretical framework attempting to unify all fundamental forces in the universe. The symmetries expressed in modular forms share remarkable similarities with certain symmetries predicted by string theory, suggesting a profound underlying connection. This demonstrates the unexpected applicability of seemingly abstract mathematical concepts to the most fundamental questions of physics, a testament to the power of number theory. It's a fascinating realm ripe with discovery; it is likely that future breakthroughs in number theory will continue to illuminate deeper understanding of our physical world. The seemingly abstract becomes strikingly concrete, revealing the unexpected unity woven into the fabric of mathematics and physics.

www.ingramcontent.com/pod-product-compliance
Ingram Content Group UK Ltd.
Pitfield, Milton Keynes, MK11 3LW, UK
UKHW020156290926
15471UKWH00005B/896

9 798349 387432